THE ALIMENTARY CANAL

MUHAMMAD LAMIN JANNEH

To order additional copies of this book, contact:

Xlibris LLC

1-888-795-4274

www.Xlibris.com

Orders@Xlibris.com

589186

Rev. date: 03/21/2014

First Edition 2014.—Xlibris LLC US; 1663 Liberty Drive Bloomington, IN 47403

To order additional copies of this book, contact:
Xlibris LLC
1-888-795-4274
www.Xlibris.com
Orders@Xlibris.com
589186

Preface

The focus of this book on the alimentary canal is to provide resources and ideas to help students and non-students of all kinds in their academic quest. Furthermore, the goal of this book is to appeal to a wide variety of students whose desire is to study and take various exams in Biology. Therefore, as a Certified Teacher (Trained Teacher), I found it necessary and appropriate to prepare educational material like this book to benefit the population in general, and in particular students in grades 6 through grade 12 in the new Gambian Educational System and other various exams in Biology around the world.

TC and HTC students in The Gambia will find this book useful during the course of their studies at The Gambia College School of Education and in their teaching careers in schools or in various professions in science generally. Personally, I was studying Mathematics with Physics as a minor for my HTC at The Gambia College School of Education when I came across Biology.

M.L. JANNEH

▊ Contents

Introduction

The Alimentary Canal is a long tube, extending from the mouth to the anus. Each part of the canal is called by a different name, and has its own characteristics, depending upon where it is located in the body. Each part performs its own particular functions. It contributes an essential part in the breaking down of food, and making it available to the body's cells. In addition to the canal itself, various organs are located along the canal's path, which aid in the process of digestion. The Alimentary Canal is essentially a conveyor belt, for the mechanical passage of food materials, allowing chemical changes to take place in the food contents along the way. The end products of the processed food is waste, while the beneficial, or good products are withdrawn, and utilized in the body.

Alimentary Canal

Before considering the canal in detail, it is useful to summarize the work, which the organs are expected to carry out. In this way, we shall be able to learn the structural arrangement, and the way in which the functions are carried out in the body.

Ingestion
This is the process of taking food into the digestive tube, which runs from the mouth to the anus.

Chewing
This takes place in the mouth, and is the process, which involves the physical breakdown of the food.

Swallowing
This is the process, which involves the active taking of the food into the throat or food pipe, and then into the stomach.

Digestion - The two types of digestion, are described as follows:

Type one - Physical Digestion - This is the actual chewing of food, done inside the mouth

Type two- Chemical Digestion - This is the process, in which enzymes produced by various glands, in the alimentary canal, act on the constituents of the food materials. The salivary glands, the stomach, and the pancreas, all produce these enzymes.

Elimination or Egestion
This is the process, in which undigested food substances are expelled, from the body, in the form of faeces.

Absorption
This is the process, by which the digested food materials passes, along the walls of the alimentary canal, and then into the circulating blood stream.

The Structure Of The Alimentary Canal

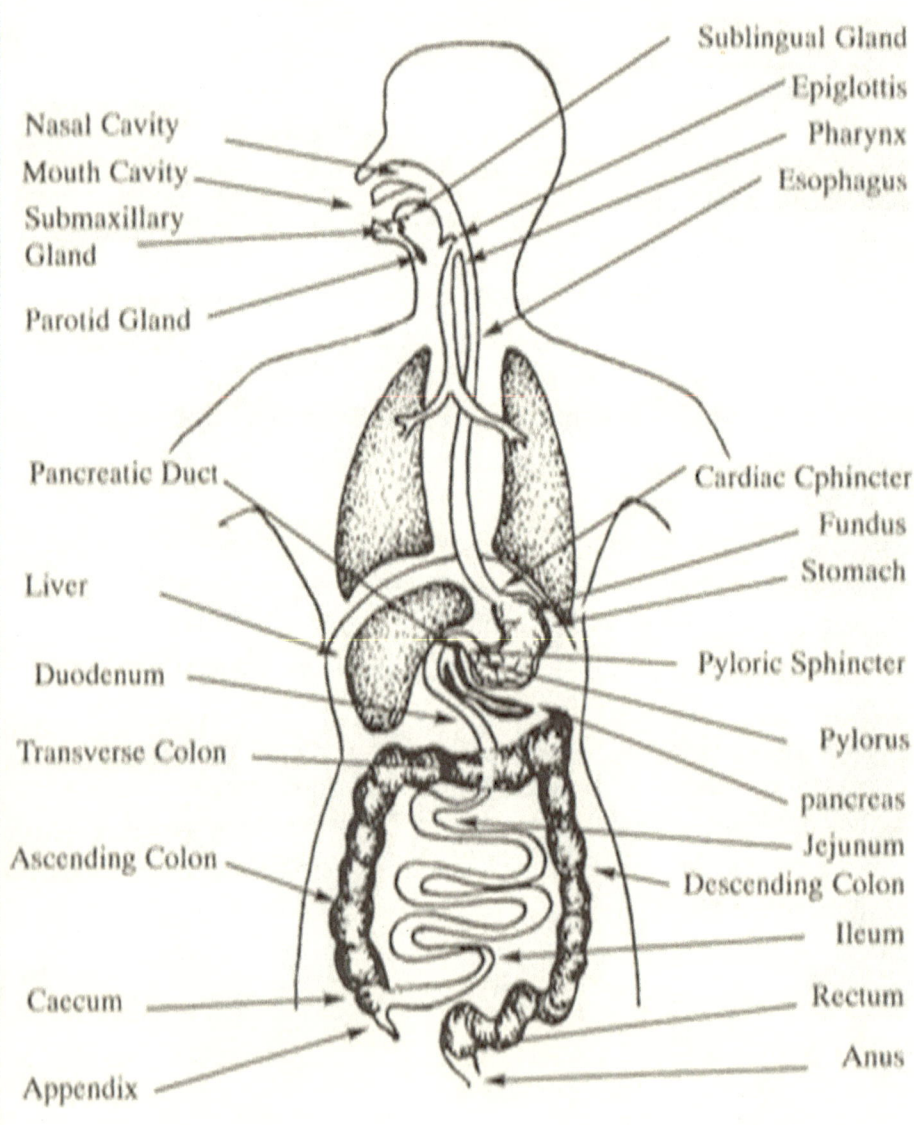

Sublingual Gland
Epiglottis
Pharynx
Esophagus

Nasal Cavity
Mouth Cavity
Submaxillary Gland
Parotid Gland

Pancreatic Duct

Liver

Duodenum

Transverse Colon

Ascending Colon

Caecum

Appendix

Cardiac Cphincter
Fundus
Stomach
Pyloric Sphincter
Pylorus
pancreas
Jejunum
Descending Colon
Ileum
Rectum
Anus

MOLAR   PREMOLAR INCISOR CANINE

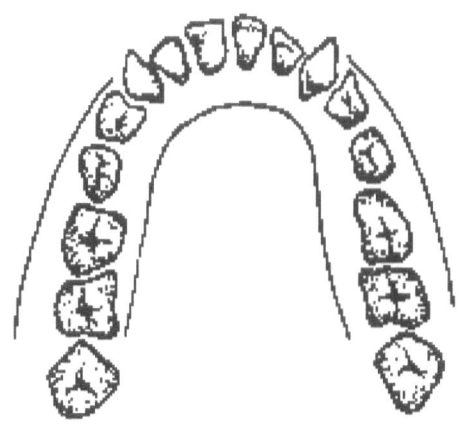

The arrangement of teeth in a human

THE STRUCTURE
OF A TOOTH

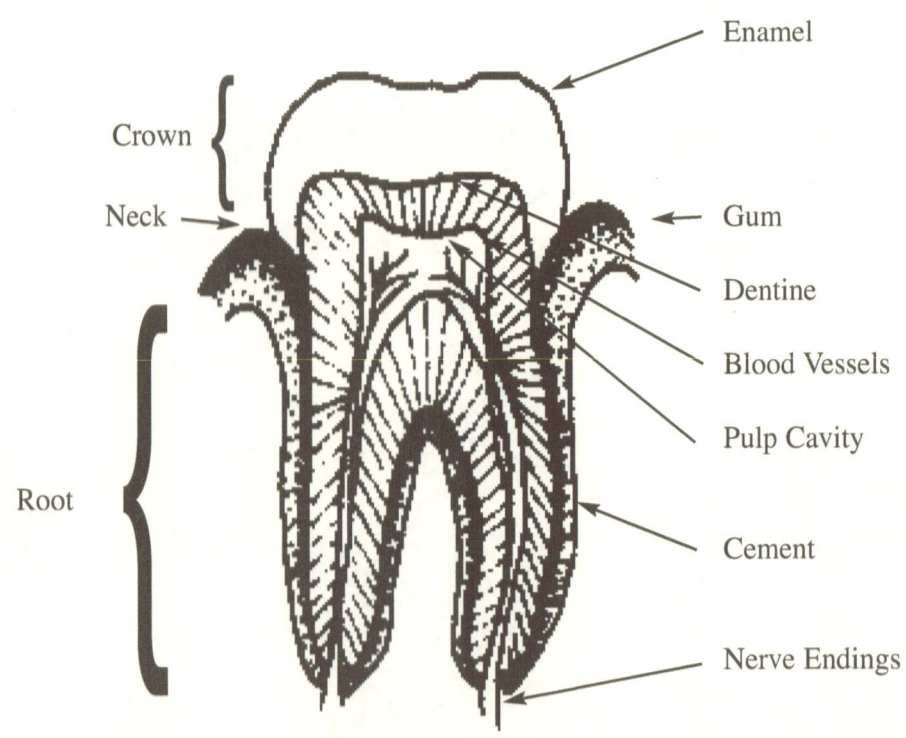

Enamel

Crown

Neck

Gum

Dentine

Blood Vessels

Pulp Cavity

Root

Cement

Nerve Endings

Ingestion

A. The Mouth

The first part of the alimentary canal is the mouth. The mouth contains the teeth, which are used for chewing. There are four types of teeth, which are named in relation to their function.

Incisors - These teeth are shaped like a chisel, and are used for cutting.
Canines - These teeth are sharp and pointed, and are used for tearing food.
Premolar - These teeth are used to crush, and to tear food.
Molars - These teeth are used for crushing, and grinding food.

A full grown human has 32 teeth; 16 on the lower jaw, and 16 on the upper.

The lower jaw, as well as the upper, each contain four incisors, two canines, four Premolar, and six molars.

B. The Structure Of The Tooth

Each tooth is made up of three separate parts:

1. **The crown-** This is the part, that is located above the gum, and physically comes in contact with food.
2. **The neck-** This is the part, surrounded by the gum.
3. **The root-** This is the part, tightly fitted into the jaw bone.

The crown of the tooth, is covered, by the hardest substance in the body, which is called enamel. Below this, is dentine, which is bone like in structure. Within the dentine, is the central pulp cavity, which is soft and contains, nerves and blood vessels. These blood vessels enter the cavity, through the root. The root is fastened into the jaw, by a material called, cementum.

THE STRUCTURAL DIAGRAM OF INGESTION

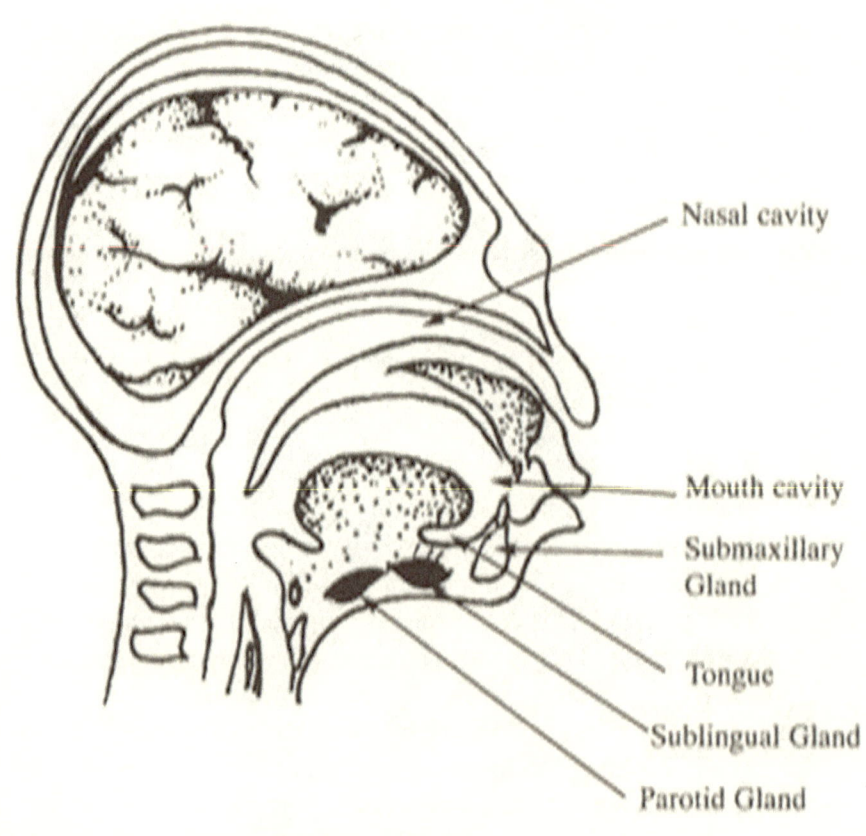

Nasal cavity

Mouth cavity

Submaxillary Gland

Tongue

Sublingual Gland

Parotid Gland

Human Head

Ingestion

C. The Salivary Glands
The salivary glands, are located in the mouth, and are accessory organs, within the alimentary canal, that are responsible for producing saliva. The three main salivary glands are as follows:

1. Parotid Gland - This is the largest of the salivary glands, and is located underneath the ear, and towards the front of the head.

2. Submaxillary Gland - This gland, is located along each side of the lower jaw.

3. Sublingual Gland - This gland is found underneath the tongue.

All these glands produce a digestive juice called salivary juice, in the mouth, and it contains amaylse which is a digestive enzyme. Amalyse breaks down food into smaller particles. Each of these glands opens into the mouth by a duct, in a tube form, that is illustrated in the diagram.

D. Throat or Pharynx
The pharynx is divided into three parts:

1. Nasal pharynx - This opens into the nasal cavity

2. Oropharynx - This opens into the mouth cavity

3. Laryngopharynx - This opens into the gullet or esophogous

The pharynx only allows taking in, and letting out of air through the mouth and nose. Food is prevented from entering this passage, by the epiglottis moving across the air passage, as it is swallowed.

THE STRUCTURAL PROCESS OF INGESTION

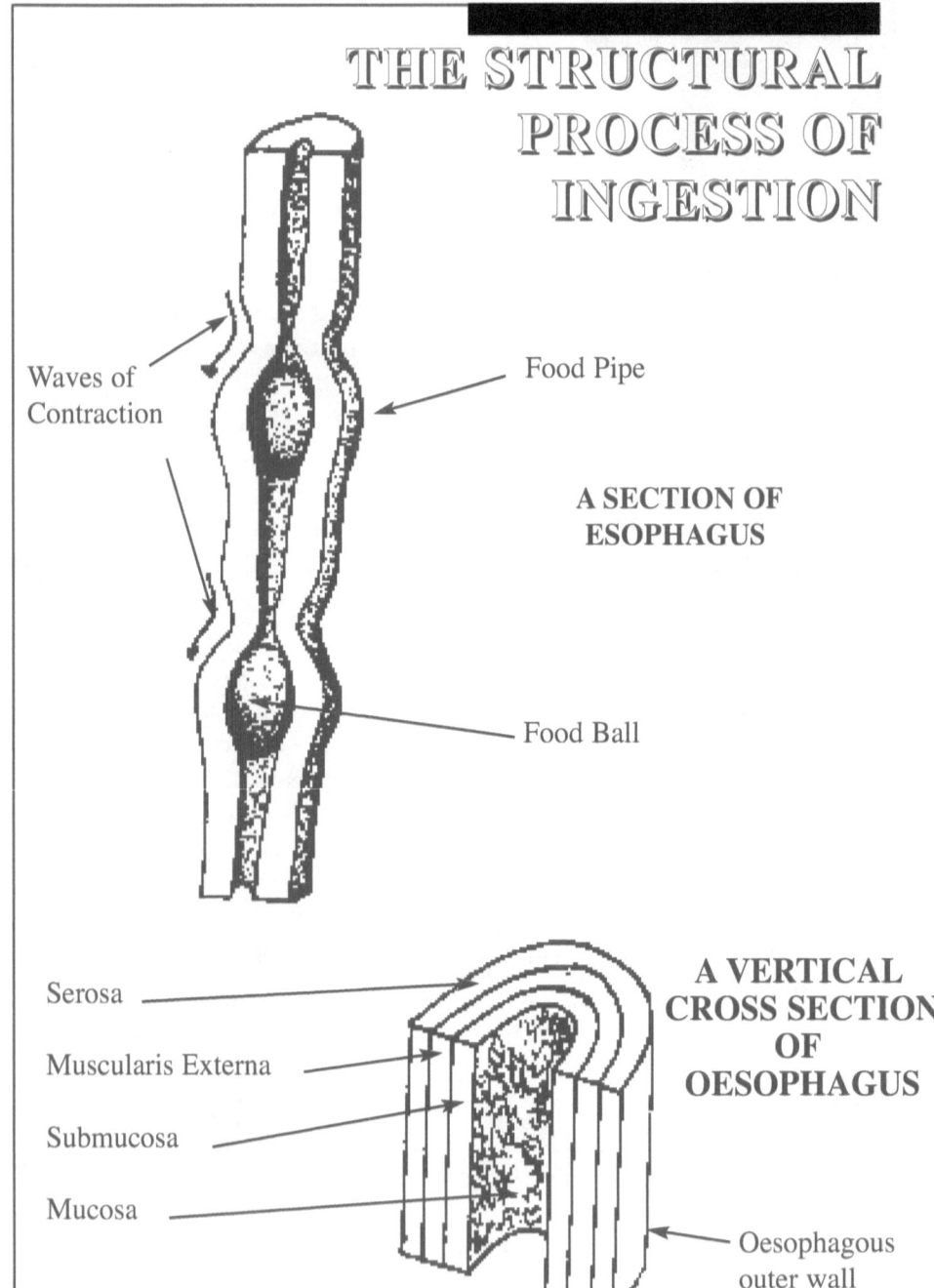

Waves of Contraction

Food Pipe

A SECTION OF ESOPHAGUS

Food Ball

A VERTICAL CROSS SECTION OF OESOPHAGUS

Serosa

Muscularis Externa

Submucosa

Mucosa

Oesophagous outer wall

Ingestion

E. Food Pipe or Esophagus

This is the tube, in which food travels on its way to the stomach. It is a long, muscular tube, and is also known as the Esophagus. The walls of the food pipe or esophagus, are muscular and it pushes solid food particles towards the stomach, by a contraction behind it. This type of movement, is known as peristalsis.

F. Internal Structure of the Esophagus

The internal structure of the esophagus, is divided into four parts, and these are:

1. Serosa
The delicate membrane of collective tissue.
2. Muscularis externa
Extenal muscular layer
3. Submucosa
A layer of loose connective tissue underlying a mucous membrane. e.g. in the wall of the intestine.
4. Mucosa
The moist layer of tissue lining in hollow organs.
e.g. alimentary canal.

A DIAGRAM OF A HUMAN STOMACH

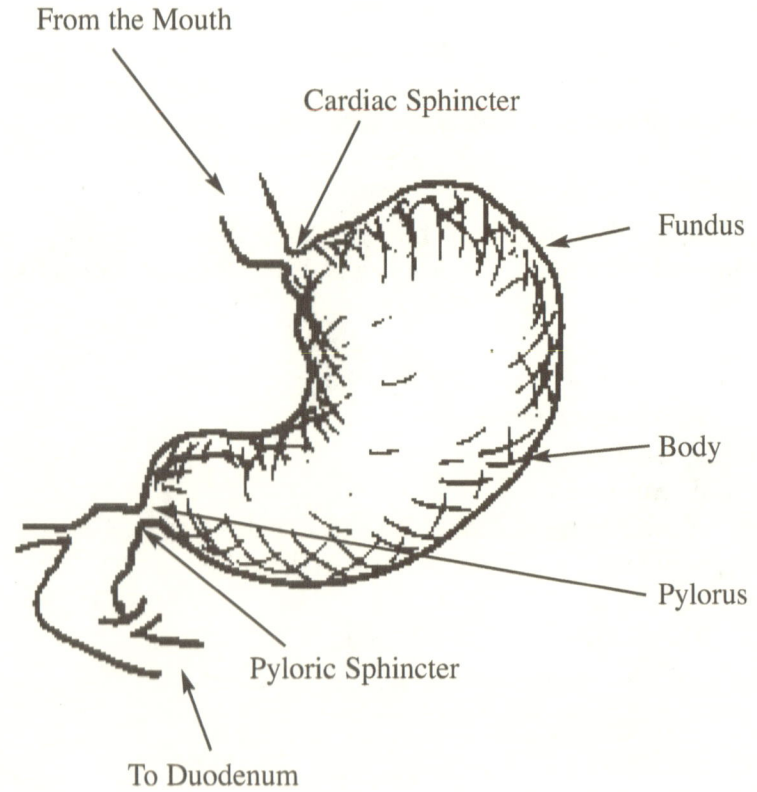

From the Mouth

Cardiac Sphincter

Fundus

Body

Pylorus

Pyloric Sphincter

To Duodenum

Digestion

G. The Stomach

The stomach has a semi curved shape, and is located immediately below the chest. It measures approximately 12 cm across, and 25 cm high. It is made up of powerful muscles, which churn food continually for proper digestion. The bottom of the stomach ends, with an opening into the small intestine.

The stomach is divided into three parts:
1. **Fundus -**The base.
2. **Body-** The main part
3. **Pylorus -** The region where the stomach joined to the duodenum.

The upper end of the stomach that opens from the esophagous, is called the Cardiac Sphincter. The lower end of the stomach, that opens into the duodenum, is called the Pyloric Sphincter. Both of these muscles, are responsible for controlling the amount of food passing through, by opening and closing. These two Sphincters act involuntarily, which means that, we do not consciously control their actions. They open and close automatically, when food is entering, and leaving the stomach.

A DIAGRAM OF A SMALL INTESTINE

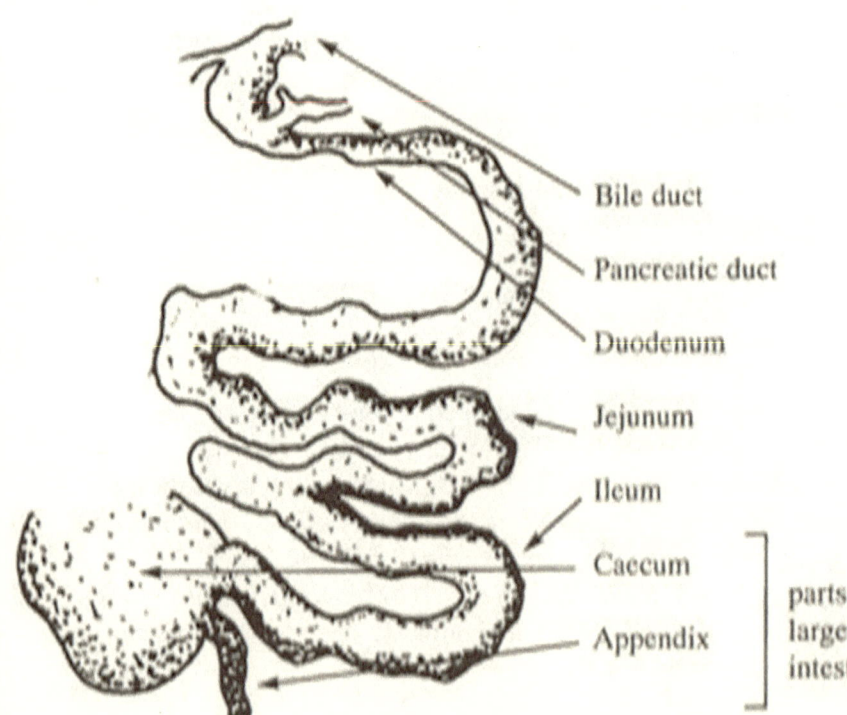

Bile duct

Pancreatic duct

Duodenum

Jejunum

Ileum

Caecum

Appendix

parts of large intestine

Digestion

H. The Small Intestine

The small intestine, is an extremely long tube, which measures over six meters. It is coiled neatly, to fit into the comparatively small space available for it, and is located directly under the stomach. Its length allows digestion, and absorption to take place very slowly, and involuntarily.

The small intestine, is divided into three parts:

1. **Duodenum** - This the first part of the small intestine measuring about approximately 30 cm long. It is a short and thick c -shaped tube, in which both the bile duct and, pancreatic duct, opens into.

2. **Jejunum** - This makes up the next two fifth of the small intestine, and it is about 2.5 meters long. It extends from the duodenum, to the ileum.

3. **Ileum** - This is the longest portion of the small intestine measuring about 3.5 meters long. It ends at, the opening into the large intestine.

Food from the stomach, first moves into the duodenum, where glands called brunnel glands, produce three types of hormones. These three types are:

1. **Secretin**- It control the rate of secretion of pancreatic enzymes .

2. **Cholecystokinin**-A hormone secreted in the upper intestinal mucosa. Active in emptying of the gall bladder mucusa.

3. **Enterocrinin**- A hormone secreted in the intestinal mucosa. It control the rate of production of digestive enzymes in the alimentary canal.

** mucosa is a membrane

A DIAGRAM OF A LARGE INTESTINE

Transverse Colon

Ascending Colon

Caecum

Appendix

Rectum

Anus

Descending Colon

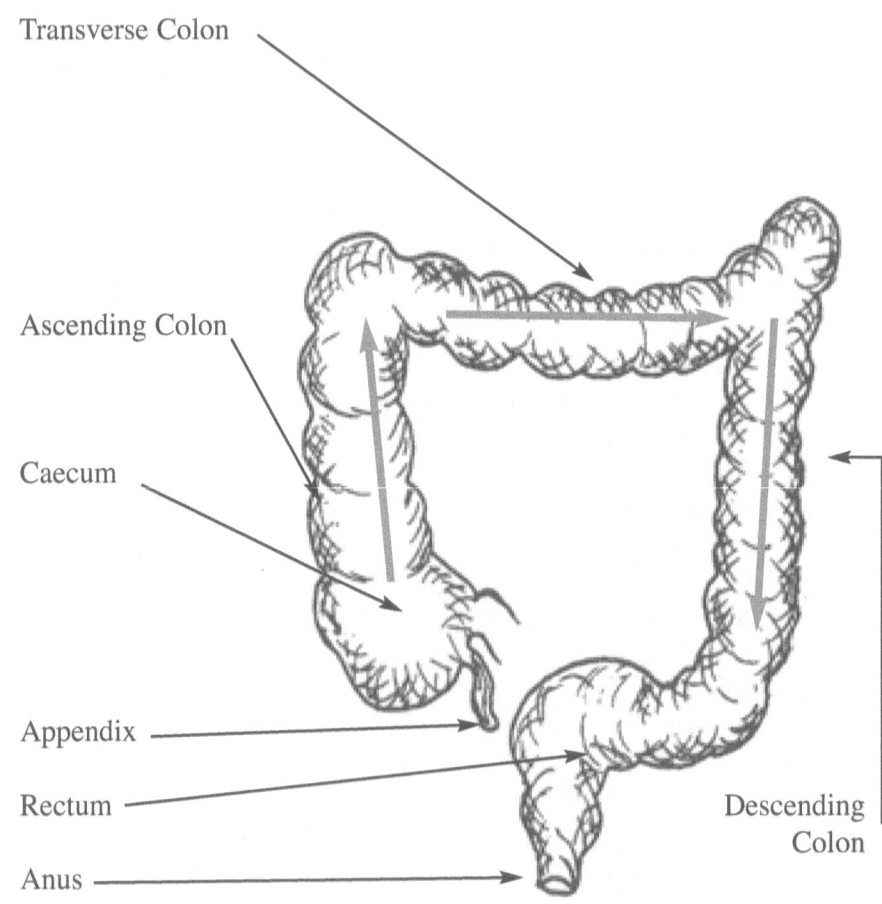

Digestion

I. The Large Intestine

The large intestine, is about 180 cm long and considerably wider, than the small intestine. It is divided into three main parts.

1. Caecum -a pouch at the junction of the ileum and colon.

2. Colon - First part of the large intestine between the ileum and rectum.

3. Rectum -The last part of large intestine in which faeces are stored and released at intervals.

CAECUM

This is a blind pouch at the lower end of the large intestine. It has three openings.

1. A wide opening into the colon.

2. An opening from the ileum.

3. Just below, is the narrow opening of the appendix, a finger-like projection at the end of the large intestine.

COLON

The colon consists of four parts:

1. Ascending colon - the material in it flows upward.

2. Transverse colon - the material in it flows laterally.

3. Descending colon - the material in it flows downwards.

A DIAGRAM OF A DIGESTIVE TRACT

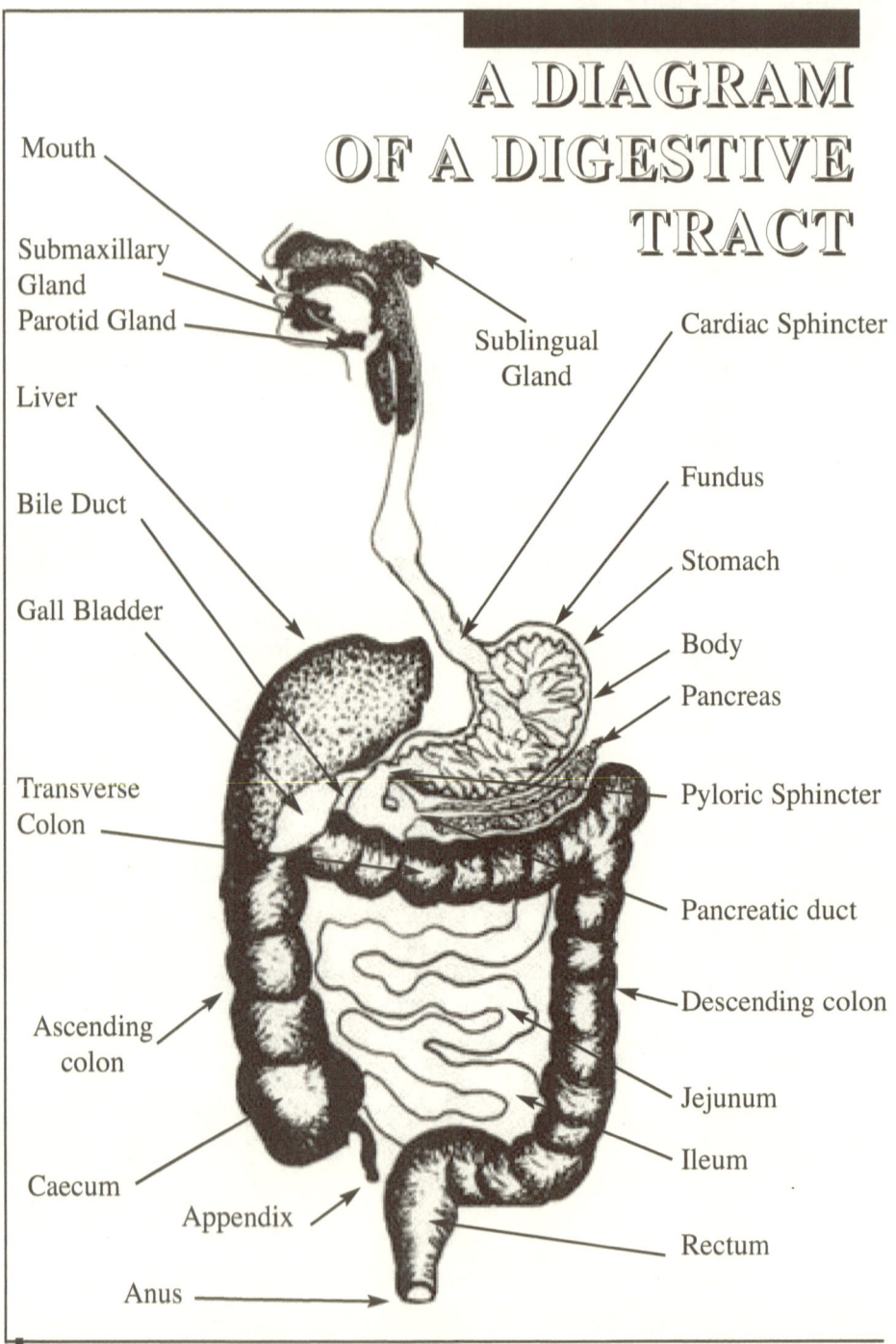

Mouth

Submaxillary Gland

Parotid Gland

Sublingual Gland

Cardiac Sphincter

Liver

Fundus

Bile Duct

Stomach

Gall Bladder

Body

Pancreas

Transverse Colon

Pyloric Sphincter

Pancreatic duct

Descending colon

Ascending colon

Jejunum

Ileum

Caecum

Appendix

Rectum

Anus

Digestion

The ascending colon, passes upwards from the caecum to the underside of the liver. It then turns left, to continues as the transverse colon, which crosses the upper part of the abdomen, from right to left. From this point, the colon bends downward, at an acute angle, to become the descending colon. This continues with the pelvics, to become the pelvic colon.

Rectum
The rectum is formed from the descending colon, and serves as a collection point for the faeces. It is the terminal point of the large intestine, and continues to the anus. Walls of the large intestine absorbs fluid materials.

PANCREAS AND LIVER

Pancreas

To the Liver and Gall
Bladder

Pancreatic duct

A DIAGRAM OF A PANCREAS

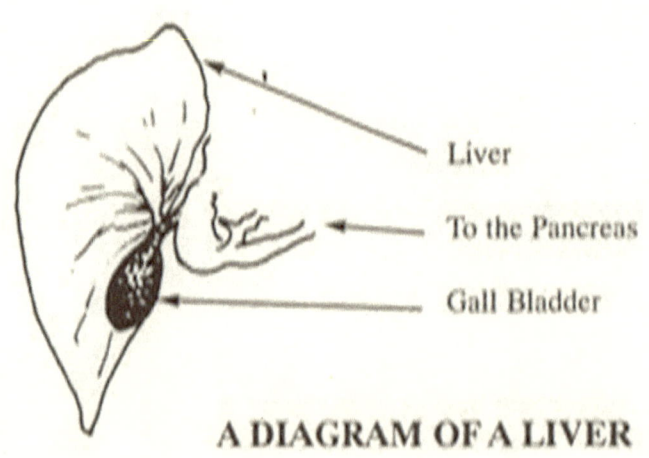

Liver

To the Pancreas

Gall Bladder

A DIAGRAM OF A LIVER

Digestion

ACCESSORY ORGANS OF DIGESTION

As indicated earlier, these organs consist of the salivary glands, the pancreas, and the liver.

PANCREAS
The pancreas is a pink colored organ about 15 cm long, and lies beneath the stomach towards the left. The pancreas function, is to produce a digestive juice, known as pancreatic juice, which contains three enzymes, and these enzymes are:

1. **Trypsinogen**
2. **Chymotrypsinogen**
3. **Enteroknase**

These enzymes are essential for the breakdown of carbohydrates, fats, and proteins, found in the food within the small intestine.

THE LIVER
The liver is the largest organ in the body, and it carries out a very important dual function. The first is that, it is responsible for keeping a balance within the body, and the second is that, it produces bile. Bile is temporarily stored in the gall bladder, which is a hollow, pear-shaped organ, attached to the underside of the liver.

GLOSSARY

ABSORBED
Means to take in.

ACIDS
A sharp tasting chemical, often in liquid form.

ADOPTION
To adjust to new circumstance, or surroundings.

ALKALINE
Any base like, that is soluble in water, and gives off ions in solution.

ALIMENTARY CANAL
The tube, running from the mouth to the anus. Digestion and absorption take place here.

ANAL SPHINCTER
The release of feces from the rectum, is controlled voluntarily by the anal sphincter.

AMINO ACIDS
Any of the nitrogenous organic acids, that form proteins necessary for all life.

AMYLASE
A starch digesting enzyme.

APPENDIX
A small, saw like appendage of the lower right portion of the large intestine.

ARTERY
A blood vessel, that transports blood away from the heart.

ASCENDING COLON
Part of the large intestine, that goes up vertically on the right side of the body.

ASCENDING
Sloping upwards.

ATOM
The smallest part of an element or the smallest unit of a matter.

BACTERIA
Micro-organisms,which have no chlorophyll, and multiply by simple division. They are harmful and useful. Useful for fermentation harmful causes diseases.

BASE
A substance, that forms salt, when it reacts with an acid.

BILE DUCT
The pipe through which, the bile pass from the liver.

BLOOD
The red fluid, circulating in the arteries and veins.

BLADDER
A bag, that can expand in size.

BLOOD VESSEL
An artery, vein or capillary.

BLOOD PLASMA
A liquid component of the blood.

BLOOD CAPILLARY
Tiny blood vessels, connecting the arteries with veins, through which the blood passes.

BOWELS
An intestines of a human being.

BOLUS
A round mass, especially a moist mass of food in the mouth, and prepared for swallowing.

BRAIN
The coordinating center of the nervous system, which in vertebrates consists of a highly organized mass of tissues, situated at the anterior end of the spinal cord, and enclosed in the bony cranium.

CAECUM
A blindly ending sac, at the junction of the small, and large intestines.

CANINES
Any of the sharp pointed teeth next to the incisors.

CAPILLARY
Any of the tiny blood vessels, connecting the arteries with veins. Fluid part of circulating blood, consisting of 91% water and 9% solids.

CAPILLARIES
The smallest type of blood vessel in the circulatory system.

CARBON
A black substance, that is formed in charcoal, or soot.

CARDIAC SPHINCTER
The opening, where the esophagus meets the stomach. It relaxes and allows food to enter the stomach, and contracts to stop food from coming up out of the stomach.

CARBOHYDRATE
An organic compound composed of carbon , hydrogen, and oxygen as a sugar or starch.

CARBON DIOXIDE
A heavy, colorless, orderless gas. It passes out, if the lungs is
 in respiration.

CAVITY
Hollow place in the tooth.

CECUM
A cavity, that opens at the end. An expanded pouch at the beginning of the large intestine.

CELLULOSE
The chief substance in the cell walls.

CELLS
The fundamental unit, of which all organisms are composed.

CELL DIVISION
The process, by which two new cells are formed, from a single parent cell.

CELL MEMBRANE
The surface layer of cells.

CEMENT
The white hardened part of the teeth.

CHEMICAL
Any substance made by a chemical process.

CHEMICAL DIGESTION
A digestion process, that takes place in the stomach, and the intestines through chemical process.

CHEWING
Biting or crushing with the teeth.

CHYMOTRYPSIN
Proteinase digestive enzyme, found on pancreatic juice, formed in the pancreas.

CIRCULATION
The movement of the blood through the arteries, and veins.

COLON
Part of the large intestine, from the caecum to the rectum, or the first part of the large intestine, where water is absorbed.

COMPOUNDS
Two or more elements, joined together to form a compound.

CONCENTRATION
An increase in density, or a proportion of molecules in a unit volume.

CONTRACTION
Becoming smaller.

CORROSIVE
To eat away.

CROWN
The exposed part of the tooth, the crest, the head of the grinding surface of the tooth.

DE-AMINATION
Removal of the nitrogenous part of an acid, prior to converting it to glycogen.

DEFECTION

To excrete waste matter from the bowls.

DEGENERATE

To lose quality.

DENTIN

The hard tissue, under the enamel of a tooth.

DELICATE

Not Strong.

DESCENDING

Directed downwards.

DIABETES

The disease, that is caused when the blood contains too much sugar, and the pancreas is not producing enough insulin.

DIAGRAM

A drawing, that explains something, as by outlining its parts.

DIGESTIBLE

That which can be digested.

DIGESTION

The process of breaking down food particles, so that they can be absorbed into he body.

DIGESTIVE ENZYMES

These are enzymes that accelerates the rate at which, insoluble compounds are broken into soluble ones. These are: amylase which act on starch, proteins which act on protein, and lipase which act on fat.

DIGESTIVE GLAND

Part of the intestine which produces digestive enzymes and in which food is digested.

DILATION

Process of widening.

DISSOLVE

To reduce into smaller particle of a substance which when mixed with water and disappears in it, is said to have dissolved.

DUCT

A tube, or a channel for passage of a liquid substance or both.

DUODENUM

The first part of the small intestine, opening from the stomach.

EGESTION

The substance, which cannot be broken down into smaller particles.

ENAMEL

The hard, white coating of teeth.

ENTEROKIN

Enzyme produced by duodenal cells, that activates trypsinogen, by cleavage of a peptide bond to produce trypsin.

ENZYME

Helps in the digestion, by breaking down food particles, that can be absorbed into the body.

EPIGLOTTIS

The thin cartilage lid, that covers the windpipe during swallowing.

EPITHELIUM

A layer of cells in an animal, lining the inside of certain organs.

ESOPHAGUS

The gullet, a tube conveying food from mouth to the stomach.

EXCRETE

To take away waste material, or waste products of chemical reactions in the cells of the body.

FECES

The undigested material, or waste products of chemical reactions in the cells of the body.

FAT

An oily or greasy material, found in animal tissue, and plant seeds.

FATTY ACIDS

Any group of organic acids in animal, or vegetable fats and oils or organic acids, containing carbon, hydrogen and oxygen only.

FERMENTATION

The breakdown of food material by yeast or bacteria, to produce energy plus carbon dioxide, and in some cases, alcohol.

FIBERS
A thread like structures, that combines with others, to form animal or vegetable tissue.

FLUID
That can flow, and change rapidly, and easily.

FOOD
Any substance, taken in by a plant or animal, to enable it to live and grow.

FOOD PIPE
The tube, through which the food passes through to the stomach.

FUNDUS
The base of an organ, or the base of the stomach.

GALL BLADDER
A sac, that contains a liquid like substance, that tastes bitter.

GASTRIC JUICE
The clear, acid digestive fluid, produced by glands in the stomach lining.

GLANDS
Any organs, that separates certain elements from the blood, and secretes them for the body use or throw off.

GLUCOSE
A simple sugar, or kind of sugar, that is found in all living cells.

GLYCOGEN
A substance in animal tissues, that is changed into glucose as the body needs it. The liver changes glycogen to glucose, and releases it into the bloodstream, to maintain healthy blood sugar level. Glucose is the body's best sugar.

GUM
The firm flesh, surrounding the base of the teeth.

HABITAT
A place in which a plant or animal lives.

HEART
Hollow muscular organ, which by rhythmic contractions, pumps blood round the body.

HEMOGLOBIN
The red, iron-containing pigment in the red blood cells, it can combine with oxygen.

HEMOPHILIA
A disease, in which the blood fails to clot.

HORMONE
A substance produced by endocrine glands into the circulatory system, that regulates the rate of bodily activities.

HYDROCHLORIC ACID
A strong, highly corrosive acid,which is a water solution of the gas, an d hydrogen chloride.

HYDROGEN
A gas, which burns rapidly, or a chemical secreted onto the blood stream on one part of the body, that controls the activity of other parts.

HYDROGEN CHLORIDE
A digestive juice, that controls the level of acid inside the stomach.

HYDROLYSIS
The addition of the hydrogen and hydroxyl ions of water to molecules.

ILEUM
The major part of the small intestine.

IMPULSES
Is a wave of motion.

INCISORS
Any of the front cutting teeth, between the canines.

INORGANIC
These are the substances, that do not have to come from a living organism. e.g. iron, salt, carbon dioxide, oxygen, etc.

INGESTION
To take food into the body.

INTESTINE
The lower part of the alimentary, extending from the stomach to the anus, and consisting of a long winding upper part (small intestine), and a shorter, thicker lower part (large intestine) bowels.

INTERNAL STRUCTURE
The inside arrangement.

INTERVALS
Once in a while, or a time in between.

JAW
Two bony parts, that hold the teeth and frame of the mouth, or two movable parts, that grasp or crush something. It bears teeth.

JEJUNUM
Part of the small intestine between duodenum and ileum in mammals. It has large villi, and is the main absorptive region.

JUICE
A liquid in or from animal tissue.

KIDNEY
A pair of organs, that separates waste products from the blood, and excretes them as urine.

LACTEAL
The tube in the center of a villus, into which passes the products of fat digestion in the intestine.

LACTIC ACID
A substance, that is produced in the breakdown of glucose during respiration.

LACTOSE
A sugar present in milk.

LARGE INTESTINE
The intestine of vertebrates, which include the caecum, colon and rectum.

LARYNX

The structure at the upper end of the trachea, containing the vocal. cords. The upper part of windpipe, which communicates with the pharynx.

LARYNGOTRACHEAL
Larynx and trachea. Chamber into which lungs open in amphibian.

LIPASE
An enzyme, which breaks down fat into fatty acids.

LIVER
The largest organ in the vertebrae animals.

LONGITUDINALLY
A diagram, placed lengthwise. Or along the length.

LUBRICATE.
To make it slippery.

LYMPH
A whitish fluid, that is derived from the blood plasma, and retuned to the circulation, via the lymphatic system.

LYMPH NODES
Where the white blood cells are produced .

LYMPHATIC
A vessel, which returns lymph from tissues, to the circulatory system.

LYMPHATIC SYSTEM
Network of fine capillaries, extending throughout the body
in the vertebrates, connected at points to the blood circulatory system.

LYMPHOCYTE
A white blood cell, which produces antibodies.

MALTOSE
Sugar produce, when starch is broken down by enzyme action.

MEMBRANE
A thin, soft layer of animal or plant tissue, that covers
or lines an organ, part, etc.

METABOLISM
Integrated network of biochemical reactions in living organisms.

MOLARS

The teeth used for grinding.

MOLECULE

Smallest particle of a substance.

MUSCLE

Contractile animal tissue involved in movement of the organism, and which also forms part of many internal organs.

MUCOUS

Secretion containing mucus.

MUSCULARIS EXTERNA

Layer of the gut wall between the sub-mucosa and serosa, consisting of a sheet of longitudinal, and sheet of circular muscles.

MUCUS

A sticky fluid, produced by animals and humans to lubricate and protect delicate surfaces, mainly stomach wall. It dilutes Hydrochloric acid inside the stomach.

NASAL

To nose.

NECK

Point of connection.

NERVES

Bundle of fibers, which carries impulses away from the cell body of neuron.

NEURON

Nerve cell, basic unit of the nervous system, specialized for the conveyance and transmission of electrical impulses.

NEUTRALIZED

That has no effect, or function.

NITROGEN

Naturally occurring.

NITROGENOUS

Containing nitrogen.

ORGAN
A group of tissues, working together to do a particular job.

ORGANIC
This commonly refers to a substance, produced by a living organism. Organic chemicals are things, like carbohydrates, protein, and fat. They're often insoluble in water.

OSMOREGULATION
The control of the quantity of water, entering and leaving the cells of an organism.

OXIDIZED
To unite with oxygen, as in burning.

PANCREAS
The gland beneath the stomach, which secrets digestive enzymes into the smallest intestine. It also produces the hormone, insulin.

PANCREATIC JUICE
A secretion containing digestive enzymes.

PANCREATIC DUCT
Carries enzymatic fluid from the pancreas.

PAROTID GLAND
Pair of salivary glands, opening into the mouth in some mammals.

PARTICLES
Tiny bits, either suspending in liquid, or in air.

PEPSIN
A stomach enzyme, aiding in the digestion of proteins.

PEPTIDES
A chemical consisting of a chin of amino acids and resulting form the partial digestion of a protein.

PEPTIDE
A chemical, consisting of a chain of amino acids, and resulting from the partial digestion of a protein.

PEPTONES
Polypeptide product of hydrolysis of proteins by enzymes such as pepsin.

PERISTALSIS
The contractions and dilation of the alimentary canal, moving the food ball downward to the stomach.

PERISTALTIC WAVE
The to and fro movement of the food along the esophagus. The alternative contraction and relaxation of the muscles, closing behind the swallowed food, and opens in front of it, combined to move liquid and solid food along the digestive tract.

PHARYNX
The area at the back of the mouth cavity, leading into the nasal cavity, gullet, and windpipe. The throat in humans, and other vertebrates.

PHAGOCYTE
A white blood cell, which can ingest foreign particles inside the body.

PIGMENT
Any coloring matter in the tissue of plants, and animals.

PLASMA
Accounts 55% of the volume of normal blood.

PLASMOLYSIS
Partial collapse of a cell, as a result of withdrawal of water by osmosis.

POISON
Substance, which in small quantity, can cause illness or death.

POISONOUS
A substance, that can injure, or kill as poison.

POLYPEPTIDE
A chain of amino acids, linked together by peptide bond.

PREMOLAR
The first molars in the teeth.

PRODUCTION
Act of producing something, or make something out.

PROJECTION
A pointed out finger-like object.

PROTEIN
This is a substance, that occurs in living matter and is essential for diet. Chemicals with large molecules, containing carbon, hydrogen, oxygen, and nitrogen.

PROTEINASE
An enzyme, which breaks down proteins.

PTYALIN
Is an enzyme in saliva, that breaks down the starch into simpler form, known as simple sugars.

PULP
Soft, sensitive tissue in the centers of a tooth.

PULP CAVITY
Is a hollow space, inside the tooth.

PYLORUS
The opening from the stomach into the duodenum.

PYLORIC SPHINCTER
The connection between the stomach, and the small intestine.

RENIN
It turns milk into curd, and they are further act on and dismantled by pepsin.

RESIDUE
That which is left, after part is taken away or the remainder.

ROOT
The embedded part of a tooth, or a hair.

SALIVA
The watery fluid, discharged by glands in the mouth, and it aids in digestion.

SALIVARY
Glands opening into or near the mouth which secrete saliva.

SALIVARY GLANDS

Glands that produce saliva in the mouth.

SECRITIN

A hormone produced by the lining of the duodenum when the acid contents of the stomach reach it. It stimulates the pancreas to produce enzymes.

SECRETION

Material or fluid which is produced and released from a cell or gland.

SEROSA

Fine membrane over external organs.

SMALL INTESTINE

The narrow section of the intestine, extending from the stomach to the large intestine. It is divided into 3 parts: duodenum, jejunum, and ileum.

SOLUBLE

That which can be dissolved in water.

STARCH

The large pouch of the intestine, between the esophagus and intestines in vertebrates.

SPHINCTER

A circle of muscle in a tube or duct. Up and down of the stomach. It regulates the movement of food through the digestive tube. To prevent backflow of partially digested food.

SUBLINGUAL

Beneath the tongue.

SUB MAXILLARY

Beneath the lower jaw.

SUB-MUCOSA

Layer of gut wall, between the mucosa and external muscular.

SWALLOWING

The act of passing from the mouth into stomach.

TEMPERATURE

The degree of hotness or coldness of anything.

TISSUE
A number of cells, that look the same in the structure and do the same function.

TONGUE
The movable, muscular structure in the mouth, used in eating, tasting, and in humans, the tongue.

TRANSVERSE
Lying across of a cross section.

TRYPSINOGEN
Digestive enzyme, found in the pancreatic juice of mammals.

UREA
A nitrogen that has chemical, and is formed in the liver from excess amino acids.

URETER
The tube along which, the urine passes from the kidney to the bladder.

URINE
A mixture of water, salt, and urea, removed from the blood by the kidneys.

VACUOLE
1. A fluid cavity, in the center of a plant cell.
2. Droplets of fluid, in the cytoplasm of animal cells.
3. Large cavity in a cell.

VEINS
Blood vessels, that carry blood to the heart.

VESTIGIAL
A degenerate part, more fully developed in an earlier stage.

VILLI
Finger-like projections, from the internal surface of the small intestine.
Finger-like objects.

VILLUS
One of thousands of Finger-like protrusions, from the internal surface of the small intestine.

TEST

TEST ONE

1.Each part of the Alimentary canal has its own characteristics, depending on what?

A. Where it is built
B. Where it is located
C. Where it is painted
D. Where it is worked

2.What best describes the Alimentary canal?

A. A conveyor belt
B. Motor
C. Liver
D. Stomach
E. Mouth

3.What is the end product of processed food?

A. Waste
B. Hot
C. Cold
D. Long
E. Over

4.What happens to the good products of the processed food?

A. Thrown away
B. Withdrawn and utilized by the body
C. Sent out
D. Put down
E. Go over

5.What is the process of taking food into the digestive system?

A. Ingestion
B. Chewing

C. Swallowing
D. Digestion
E. Absorption

6.What is the name of the process by which undigested food substance is expelled from the body?

A. Chewing
B. Throwing out
C. Egestion
D. Swallowing
E. Digestion

7.What is the actual chewing of food inside the mouth?

A. Physical digestion
B. Smooth digestion
C. Rough digestion
D. Colored digestion
E. Below digestion

8.What is the name of the process which involves the physical break down of the food?

A. Sitting
B. Jumping
C. Chewing
D. Ingestion
E. Swallow

9.One of the following does not produce enzymes?

A. Salivary glands
B. The stomach
C. Pancreas
D. Appendix

Use this diagram to answer questions 10 to 14.

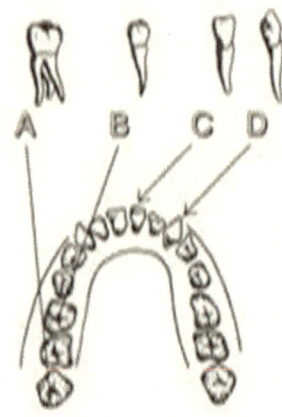

10. What does A stand for ?

   A. Premolar
   B. Molar
   C. Incisor
   D. Canine
   E. Mouth

11. What does arrow C represent?

   A. Canine
   B. Incisor
   C. Premolar
   D. Molar
   E. Lips

12. What does arrow D represent?

   A. Molar
   B. Premolar
   C. Incisor
   D. Canine
   E. Lumbar

13. What does arrow B represent?

A. Canine
B. Premolar
C. Molar
D. Incisor
E. Head

14. Human Teeth consist of how many types of teeth?

A. Molar, Premolar, Incisor
B. Molar, Premolar, Incisor, Canine
C. Premolar and Incisor
D. Molars
E. Premolars

15. In the arrangement of teeth the human incisors can be found:

A. Front
B. Back
C. Below
D. Medium
E. Molar

Use this diagram to answer questions 16 to 20.

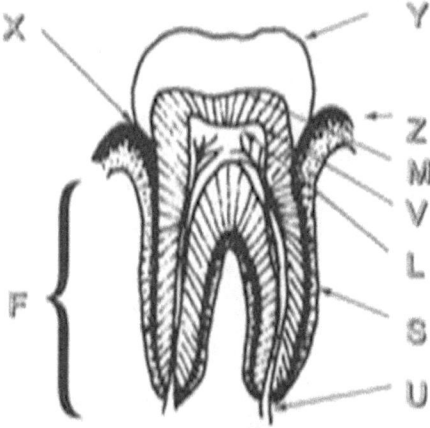

16.What does arrow F represent?

A. Gum
B. Root
C. Enamel
D. Neck
E. Cement

17.What does arrow Z represent?

A. Gum
B. Enamel
C. Pulp cavity
D. Root
E. Neck

18.What does arrow M represent?

A. Gum
B. Dentine
C. Neck
D. Root
E. Cement

19. What does arrow S represent?

A. Cement
B. Nerve endings
C. Root
D. Dentine
E. Gum

20.What does arrow L represents?

A. Root
B. Pulp cavity
C. Dentine
D. Root
E. Neck

1.The first part of the Alimentary canal is called:

A. Mouth
B. Toe
C. Back
D. Hand
E. Head

2.One of the following can be found in a mouth:

A. Stone
B. Wood
C. Nails
D. Teeth
E. Ear

3.These teeth are best described as shaped like a chisel:

A. Incisors
B. Knife
C. Canines
D. Premolar
E. Molars

4.One of the following in the mouth is used to crush and tear food:

A. Molars
B. Incisors
C. Canines
D. Premolar
E. Knife

5.The full grown man has how many teeth?

A. 48
B. 40
C. 24
D. 32
E. 16

6.How many incisors are there in both lower and upper jaw combined?

A. 5
B. 4
C. 6
D. 2
E. 8

7.How many premolars are there in both lower and upper jaw combined?

A. 10
B. 8
C. 4
D. 2
E. 6

8.How many molars are there in both lower and upper jaw combined?

A. 12
B. 10
C. 7
D. 8
E. 9

9.How may canines are there in both lower and upper jaw combined?

A. 4
B. 6
C. 5
D. 7
E. 9

10.The crown of the tooth is located where?

A. Top
B. Above the gum
C. Stomach
D. Nail
E. Hand

11. The hardest substance in the body that covered the crown of the tooth is called:

A. Enamel
B. Cement
C. Neck
D. Below
E. Upper

12. The central pulp cavity of the tooth can be found:

A. Below the head.
B. Within the dentine
C. Top
D. Below
E. Hair

13. The root is fastened into the jaw by a material called:

A. Block
B. Cementum
C. Wood
D. Steel
E. Brass

14. Salivary glands can be found in the:

A. Legs
B. Mouth
C. Hair
D. Finger
E. Computer

15. What does the salivary glands produce?

A. Blood
B. Milk
C. Saliva
D. Meat
E. Nails

16. The largest salivary gland is called:

A. Parotid Gland
B. Submaxillary Gland
C. Protein
D. Pancreas
E. Teeth

17. Sublingual Gland can be found in one of the following places in the mouth:

A. Underneath the tongue
B. In the liver
C. At the hand
D. In the eye
E. In the stomach

18. Parotid, submaxillary and sublingual glands produce a digestice juice called:

A. Salivary juice
B. Blood
C. Milk
D. Water
E. Fluid

19. The pharynx is divided into how many parts?

A. 4
B. 6
C. 5
D. 3
E. 10

20. Where does oropharynx open into?

A. The mouth cavity
B. The stomach
C. The box
D. The field
E. The hollow

TEST THREE

1.The pharynx conducts one of the following:

A Taking in of air and letting out of air
B. Taking in of air and letting out of blood
C. Taking in of are and letting out of of saliva
D. Taking in of water and letting out of CO2.
E. Taking in of wood and letting out of water

2.The highway for the food is called

A. Food pipe or esophagus
B. Ducter
C. Stomach
D. Hand
E. Finger

3.The movement of food in the food pipe is called:

A. Persistalsis
B. Jumping
C. Sleeping
D. Standing
E. Laying

4.One of the following is not part of the internal structure of the esophagus

A. Parotid
B. Serosa or Adrentia
C. Muscularis externa
D. Submucosa
E. Mucosa

Use this diagram to answer questions 5 to 11.

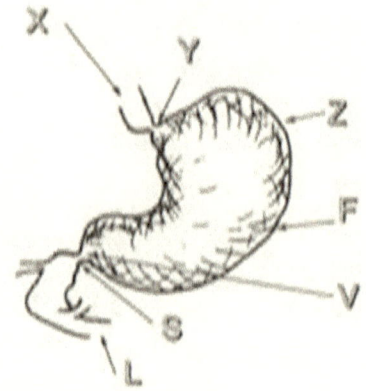

5. What does arrow X represents?

    A. From the mouth
    B. Plyorus
    C. Fundus
    D. Neck
    E. Toe

6. What does arrow L represent?

    A. Neck
    B. Cardiac Sphincter
    C. Body
    D. Mouth
    E. To Duodenum

7. What does arrow V represent?

    A. Plyorus
    B. Fundus
    C. Cardiac Sphincter
    D. Mouth
    E. Body

8. What does arrow S represent?

A. Plyoric Sphincter
B. Body
C. Fundus

9.What does arrow Y represent?

A. Cardiac Sphincter
B. Fundus
C. Body
D. Plyorus
E. Duodenum

10.What does arrow F represent?

A. Body
B. Fundus
C. Mouth
D. Plyorus
E. Neck

11.What does arrow Z represent?

A. Body
B. Fundus
C. Plyorus
D. Duodenum
E. Cardiac Sphincter

12.The stomach is best described as:

A. Basket ball
B. A curve like ball
C. A semi curved shape
D. Rectangle
E. Square

13. What is the approximate length of the stomach across?

A. 110 cm.
B. 12 cm.

C. 20 cm.
D. 36 cm.
E. 7 cm

14. What is the height of the stomach?

A. 35 cm.
B. 2.5 cm
C. 25 cm
D. 25 m
E. 50 m

15. The stomach is divided into how many ports?

A. 2
B. 4
C. 7
D. 3
E. 6

16. The base of the stomach is called:

A. Ball
B. Fundus
C. Body
D. Plyorus
E. Gear

17. The upper end of the stomach that opens from the esophagus is called:

A. Longitudinal
B. Esophagus
C. Buttom
D. The Cardiac Sphincter

18.What does the plyoric sphincter represent?

A. The lower end
B. The middle end
C. The bottom end
D. The far end
E. The below end

19.Two of the muscles that are responsible for controlling the amount of food passing through by opening and closing are called:

A. Pyloric and Butter sphincter
B. Pyloric and Toe
C. Pyloric and cardiac sphincter
D. Taste Food
E. Gunjur and Bakau

20.Pyloric and cardiac sphincter opens and closes:

A. Fastly
B. Slowly
C. Automatically
D. Rapidly
E. Upperly

TEST FOUR

Use this diagram to answer questions 1 to 5.

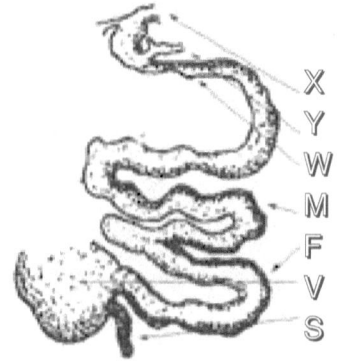

1.What does arrow V represent?

A. Caecum
B. Body
C. Bile Duct
D. Jejunum
E. Duodenum

2.What does arrow M represent?

A. Ileum
B. Body
C. Jejunum
D. Bile duct
E. Appendix

3.What does arrow S represent?

A. Appendix
B. Ileum
C. Duodenum
D. Bile duct
E. Caecum

4.What does arrow V and S represent?

A. Large intestine
B. Ileum
C. Appendix
D. Pancreatic Duct
E. Caecum

5.What does arrow Y represents?

A. Pancreatic Duct
B. Bile Duct
C. Ileum
D. Caecum
E. Jejunum

6.Small intestine measures over:

A. 27m
B. 3m
C. 6m
D. 1m
E. 4m

7.The small intestine is divided in to how many parts?

A. 4
B. 2
C. 3
D. 5
E. 6

8. The first part of the small intestine is called:

A. Jejunum
B. Duodenum
C. Ileum
D. Stomach
E. Head

9.The longest portion of the small intestine is called:

A. Ileum
B. Jejunum
C. Stomach
D. Duodenum
E. Pancreas

10.Jejunum measures about:

A. 3m
B. 2.5m
C. 7m
D. 9m
E. 10m

11.Ileum measures about:

A. 3.5m
B. 4 m
C. 7m
D. 8m
E. 9m

12.Duodenum measures about:

A. 40m
B. 50m
C. 20m
D. 30m
E. 10m

13.Where does the food from the stomach first move into?

A. Jejunum
B. Large intestine
C. Duodenum
D. Kidney
E. Pancreas

14. The brunnel glands can be found in?

A. Stomach
B. Duodenum
C. Kidney
D. Liver
E. Pancreas

15. The brunnel glands produced three types of hormones called:

A. Secretin, cholecystokinin, liver
B. Kidney, Pancreas, Pancake
C. Secretin, Cholecystokinin, Enterocrinin
D. Gunjur, Serekunda, Banjul
E. Pancreas, Liver, Stomach

Use this diagram to answer questions 16 to 20.

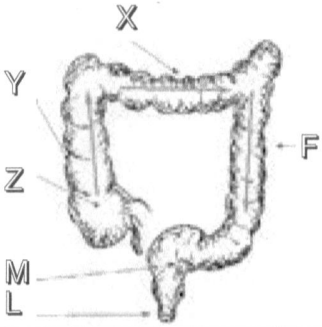

16.What does arrow F represent?

A. Ascending colon
B. Descending colon
C. Rectum
D. Anus
E. Caecum

17.What does the arrow L represent?

A. Rectum
B. Anus
C. Ascending colon
D. Transverse colon
E. Caecum

18.What does the arrow Z represent?

A. Caecum
B. Rectum
C. Ascending colon
D. Descending colon
E. Anus

19.What does the arrow M represent?

A. Anus
B. Rectum
C. Ascending colon
D. Transverse colon
E. Descending

20.What does the arrow X represent?

A. Transverse colon
B. Descending colon
C. Anus
D. Rectum
E. Caecum

1.What is the length of the large intestine?

    A. 200 m
    B. 180m
    C. 60 m
    D. 70m
    E. 90m

2.The three parts of large intestine are called:

    A. Caecum, colon, stomach
    B. Colon, blue, yellow
    C. Rectum, colon, stomach
    D. Star, colon, legs
    E. Caecum, colon, sleep

3.The blind pouch at the lower end of the large intestine is called:

    A. Colon
    B. Caecum
    C. Rectum
    D. Hand
    E. Hoe

4.Which of the following three consists of colon:

    A. Descending, transverse, below
    B. Descending, bending, ascending
    C. Ascending, transverse, descending
    D. Below, up, down
    E. Come, true, Hawa

5.The ascending colon, the material in it flows:

    A. Downwards
    B. Upwards
    C. Middle wards
    D. Medium words
    E. Length down

6.In the descending colon, the material in it flows:

A. Downwards
B. Upwards
C. Levelwards
D. Medium wards
E. Semi wards

7.In the transverse colon, the material in it flows:

A. Medium
B. Laterally
C. Below
D. Rear
E. Near

8.A finger like projection at the end of the large intestine could be one of the following:

A. Kidneys
B. Appendix
C. Meat
D. Pancreas
E. Liver

9.Rectum is formed from one of the following:

A. Descending colon
B. Ascending colon
C. Stomach
D. Kidney
E. Liver

10.Walls of the large intestine absorb only:

A. Fluid materials
B. Boxes
C. Stones
D. Proteins
E. Feces

Use this diagram to answer questions 11 to 16.

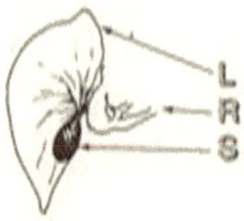

11. What does the arrow W represent?

A. Pancreatic duct
B. Pancreas
C. Liver
D. Bladder
E. Neck

12. What does the arrow L represent?

A. Liver
B. Pancreas
C. Gall bladder
D. Pancreatic duct
E. Neck

13 What does the arow R represent?

A. To the bladder

B. To the pancreas
C. Liver
D. Gall bladder
E. Mouth

14.What does the arrow S represent?

A. Gall bladder
B. Liver
C. To the pancreas
D. Parotid
E. Tongue

15.What does the arrow X represent?

A. To the liver and gall bladder
B. Liver
C. Gall bladder
D. pancreatic duct
E. Intestine

16. What does the arrow F represent?

A. Liver
B. Kidney
C. Intestine
D. Pancreas
E. Toe

17.The three accessory organs of digestion are:

A. Kidney, Liver, and Stomach
B. Salivary gland, Pancreas, and Liver
C. Head, mouth , and legs
D. Stomach, Hands, and Neck
E. Legs, Blue, and Back

18.The three enzymes from the pancreas that are responsible for the breakdown
of carbohydrates, fats and proteins are:

A. trypsinogen, chymotrypsinogen and liver
B. Enteroknase, chymotrypsinogen, and trypsinogen
C. enteroknase, kidney, and blood
D. trypsinogen, kidney and hands
E.   Stomach, Gunjur. and Bakau

19.The teeth in the mouth are used for:

A. Laughing
B. Looking
C. Chewing
D. Sitting
E. Following

20.The blood vessel in the tooth enter the pulp cavity through the"

A. Road
B. Bar
C. Bone
D. Upper
E. Root